The Environmental ABC's

The Environmental ABC's

By Aja -Fatou Jagne

Illustrated by Sophisticated Press

Publisher: Aja-Fatou Jagne

Publishing Consultant:

SP
SOPHISTICATED
PRESS

Dedication

This Alphabet Book is dedicated to one of the most influential people in my life- my Grandmother Ndey Amie Taal. Although my Grandmother was not awarded the privilege of a formal education, she was the smartest and most resourceful person that I've been blessed to know and have as an integral part of my life for the past 3 decades. Advancing my education made my Grandmother so proud of me, and I hope to spark an interest in the lives of the next generation through this educational tool. Thank you Grandma for showing me the importance and fostering my education from a very early age. You will forever be in our hearts, may your beautiful soul rest in eternal peace.

This book is also dedicated to my three wonderful God-children: Alisandra Ndioba Fall, Ahmad Amari McCauley and Luna Alexandria Blackman. I hope that they are inspired by my Environmental ABC's and that it sparks an interest in how the air we breathe, the water we drink and the food we eat all affects our well-being. God mommy loves you so much and will always be in your corner.

"It is up to us to keep and preserve our beautiful Earth and to keep our environment clean and safe, so that the next generation may benefit from it the way we have. Small actions make such a big impact when it comes to our environment."

- Aja Jagne

A

Air: Invisible gas that surrounds the Earth.

B

Bacteria: Germs that are so small, they can only be seen through a microscope.

C

Climate Change: The change in weather patterns in a region of the Earth, over a long period of time.

D

Disease: An illness or sickness.

EXAM ROOM 1

E

Environment: Everything around us, including the air, soil, water, plants, and animals.

F

Food: The material that people
and animals eat to live and grow.

AUTO OFF 165F ⊙⊙⊙

AUTO OFF 155F ⊙⊙⊙

AUTO OFF 145F ⊙⊙⊙

G

Gas: A state of matter that has no shape, it takes the shape of its container.

H

Hazard: Any source of potential harm to something or someone.

I

Inspection: A careful look at a business, service, or material.

J

Justice: Fair treatment of everyone regardless of race, color, or income.

K

Kit: A set of parts for building or testing something.

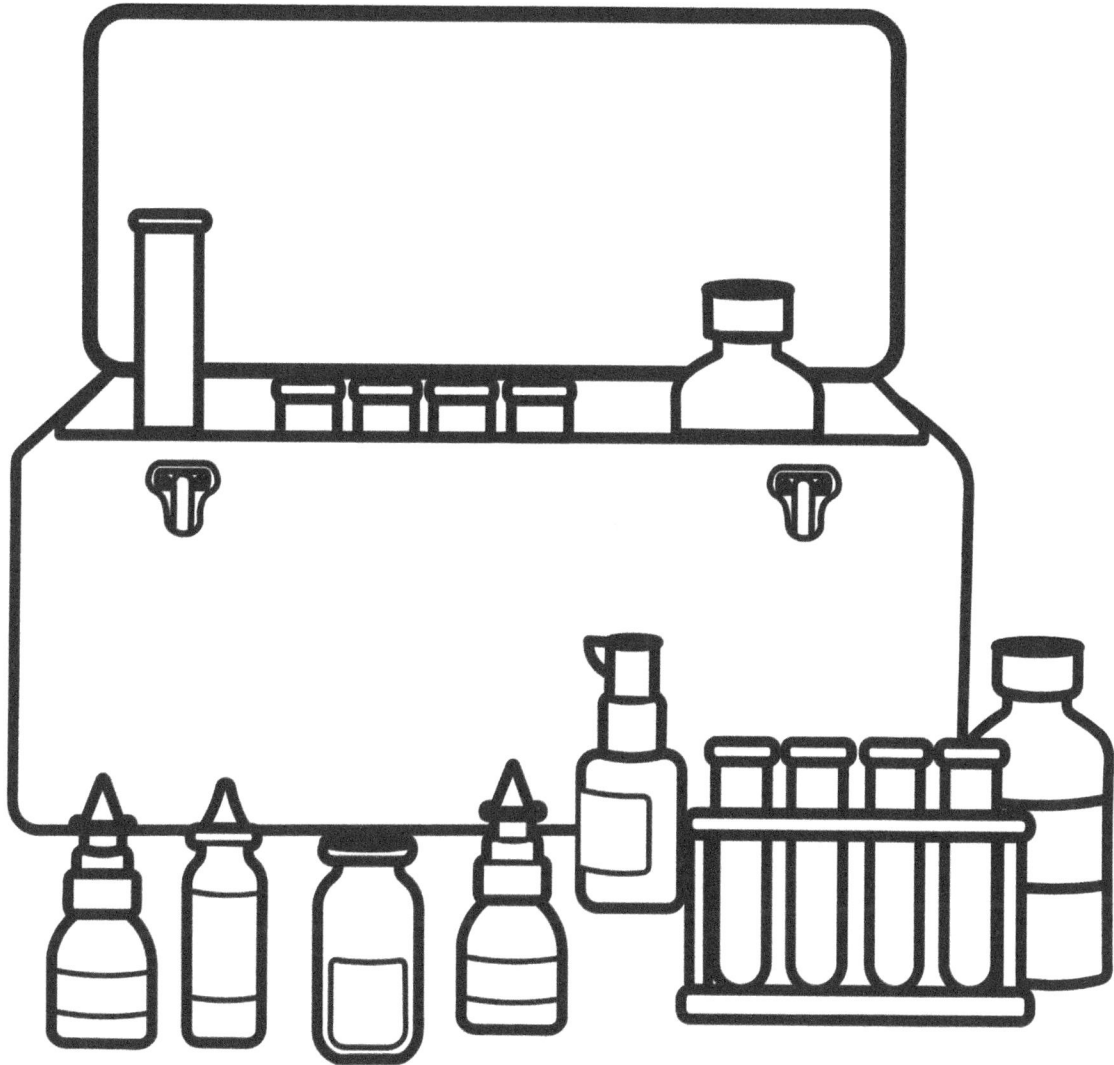

L

Lead: A soft, silvery-white, or grayish metal that can make your water brown or discolored.

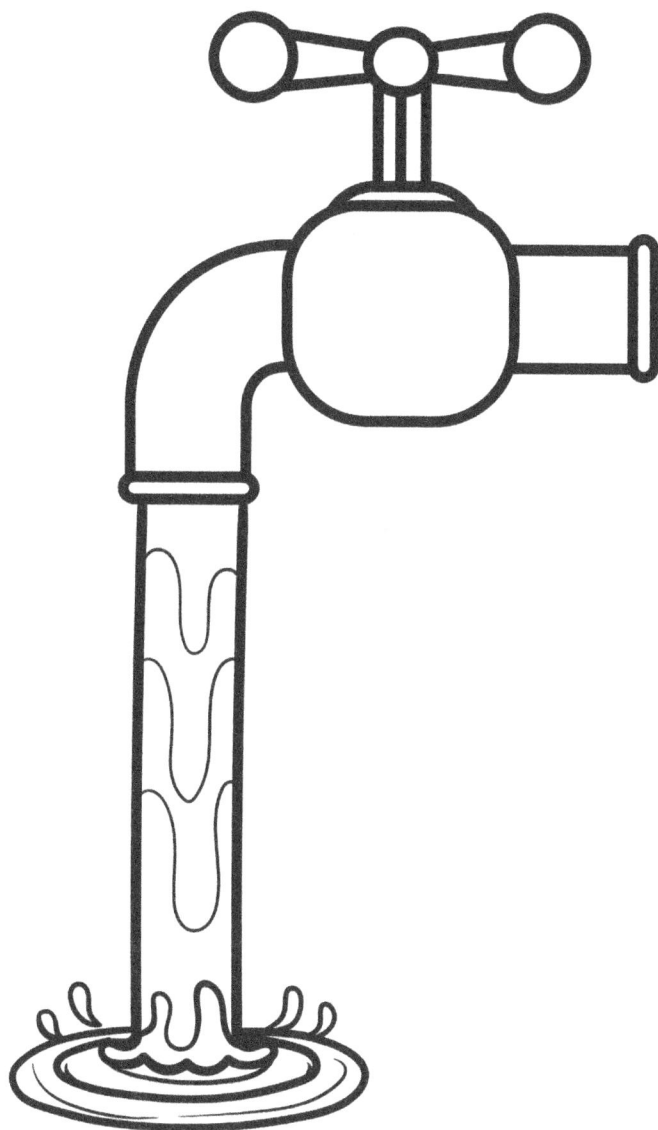

M

Mask: A face covering that stops droplets from the mouth and nose from spreading in the air.

N

Noise: Unwanted sound that may be unpleasant, loud, or make it hard to hear.

O

Ozone: A gas that has no color or odor, and is made up of three oxygen atoms.

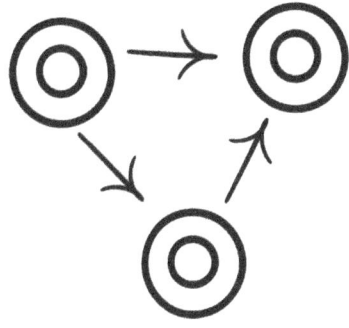

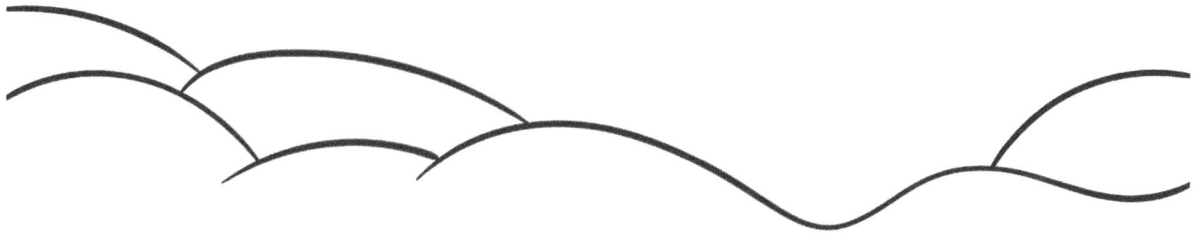

P

Pollution: When the environment is contaminated, or dirtied, by waste, chemicals, and other harmful things.

Q

Quarantine: Separating people, animals, and things to stop disease from spreading.

R

Recycle: To use something that was used before for a different purpose.

S

Smog: A type of air pollution that makes the air very hard to see through.

T

Toxin: Something that is poisonous and can make a person or animal very sick.

U

Ultraviolet (UV) light: A type of radiation with wave lengths that is shorter than the violet end of light.

V

Vapor: A spread out matter like smoke or fog that floats in the air and makes the air less clear.

W

Water: A clear liquid with no odor, that makes up the sea, lakes, rivers, and rain. It also makes up a lot of the human body.

X

X-Ray: A type of radiation that creates pictures of the inside of your body.

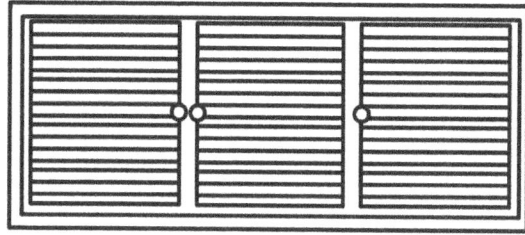

Y

You: Any person in general.

Z

Zoonosis: A disease that can be transmitted to humans from animals.

About The Author

Aja-Fatou Jagne is an Environmental Specialist. Aja was born in Atlanta, Georgia and raised in The Gambia, West Africa, where both of her parents originated. She earned a Bachelor's degree in Environmental Health Science from Benedict College in 2013 and a Master's in Public Health with a concentration in Environmental Health from Georgia State University in 2019. Aja prides herself on promoting public health, specifically through environmental activism and awareness.

www.ingramcontent.com/pod-product-compliance
Lightning Source LLC
Chambersburg PA
CBHW080630030426
42336CB00018B/3140